PHOTOGRAPHIC SURVEILLANCE TECHNIQUES

For Law Enforcement Agencies

Kodak

SURVEILLANCE

WHO?
Suspected Criminals
Rioters
Demonstrators
Crowds

WHAT?
Criminal Acts
Violent Behavior
Arrests

WHEN?
Day
Night
Anytime

WHERE?
City Streets
Public Buildings
Parks
Nightclubs
Anywhere

WHY?
Identification
Aid to Conviction
Record
Training

HOW?
Telephotography
Infrared Photography
Low-Light Photography
Time-Lapse Photography
Motion-Picture Photography

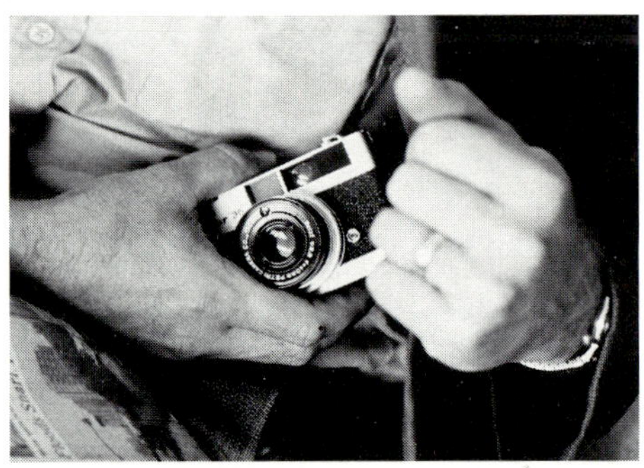

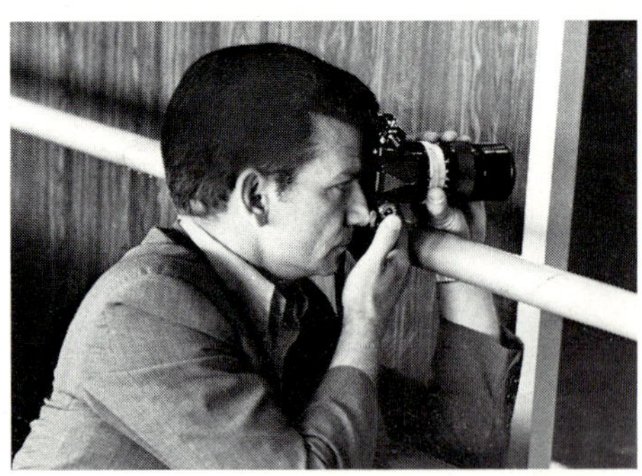

[redacted]

PHOTOGRAPHIC SURVEILLANCE TECHNIQUES
For Law Enforcement Agencies

TABLE OF CONTENTS

This pamphlet is designed to provide information and guidance to those persons in law enforcement, government, and industry who are concerned with surveillance techniques. The pamphlet does not attempt to cover all possible techniques or systems for accomplishing desired objectives. Persons interested in surveillance techniques should, of course, consult federal, state, and local laws to determine the methods which can properly be applied in their jurisdictions.

Standard Book Number 0-87985-050-7

INTRODUCTION

Surveillance is the general term used to describe the observation of suspected persons. Its purpose is to establish the identity of a person and to record illegal activities. The camera is an excellent means of recording such information.

Photography under adverse lighting, photography without detection, photography at long distances—all are required of the police photographer assigned to make surveillance records. Faced with these problems, the police photographer must make a suitable choice of camera, camera position, film, and other equipment in order to get the photographs required.

Successful surveillance requires good police work. Good surveillance photographs require good police work *plus* good photographic techniques. Such techniques include the use of telephoto lenses, the use of infrared flash or lamps, the processing of films for maximum film speed, and the use of motion-picture and time-lapse cameras.

Observation of a person or group of persons may be active, that is, a camera is pointed, focused, and controlled continuously by an operator. Or it may be passive, the camera being triggered by an action of the suspect or used as a continuous or intermittent recording device.

Active surveillance may be overt or covert. In observation of demonstrations or fire scenes, there may be little need or opportunity to hide the camera or operator. Where observation of suspected criminal activity is concerned, however, concealment of the photographer and cameras may be vital to obtain the required evidence.

Passive surveillance relies upon proper setup of camera traps or installation of continuous or intermittent surveillance cameras to capture the required pictures. Without operator control, initial, detailed analysis of the photographic detection system must be done to provide, with increased reliability, a record of the required information or identification.

SURVEILLANCE, EVALUATING THE TASK

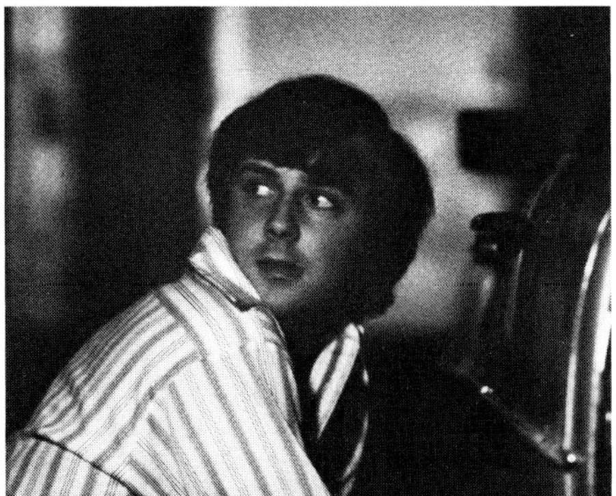

FIG. 1: A clear image of the subject's face is all important in establishing identification.

Picture Requirements

Before attempting any photographic stakeout, decide what photographs will be relevant to the investigation and will contribute to identification of a suspect, show his actions, or yield other useful evidence. Without such planning, you may get photographs which prove to have no value for the investigation or subsequent prosecution.

Where only identification is required, the most important aspect of the picture will be the subject's face. Other attributes of an individual or group may also be useful for complete and unambiguous identification. Mode of dress, mannerisms, or unusual body characteristics may lead to a positive identification where the face of a suspect is obscured or not sufficiently clear.

In addition to positive identification which establishes the presence of a suspect at a particular place at a given time, a record of the suspect's actions can help to establish that a crime has been committed. To establish the facts, record the actions of the suspect with a series of still pictures or use a motion-picture camera for a continuous record. Attempt to photograph objects in the hands of the suspect or objects being passed from the individual to another. Weapons, tools, or other objects may connect the individual with the commission of a crime. Show incriminating actions such as forcible entry, passing of contraband or illegal goods, taking bets or bribes, and sale of drugs.

Subject Position

Normally, you will have no control over the position of the subject. However, if you have studied the subject's habits, you may have a choice of site at which you can photograph him and his activities. Anticipation of the subject's movements and position may provide the only opportunity to get useful photographs.

Try to get photographs when there is a natural pause in a subject's activities such as opening a door, greeting an acquaintance, or waiting to cross the street. Remember that identity can be established in one picture and activities recorded in related photographs.

Public areas are most accessible for surveillance purposes, but they may also be the most difficult places to watch. You will have many extraneous activities which may have no relationship to the suspect, but you must be alert to their possible involvement. Crowds, traffic, or parked cars may block your view of the suspect's position.

Surveillance in controlled areas such as parking lots, race tracks, building lobbies, or sports arenas may give you more opportunity to define the suspect's likely position. You can enlist the aid of attendants, janitors, or building superintendents, where necessary, to inspect the area for surveillance. In some areas, you may even have the opportunity to supplement the existing lighting without arousing the suspicion of the subject.

Lighting

One of the principal problems that you will encounter in making surveillance photographs is the level of illumination. Most criminal suspects will try to cloak their movements and actions. Darkness is one of the most useful assets for such a suspect.

With fast films, infrared films and flash, or low-light-level scopes, the surveillance photographer can overcome many of the limitations imposed by darkness. The lighting on the subject will govern, to a large degree, the choice of film and camera equipment for a surveillance task. The use of a high-speed film or infrared film can be a relatively inexpensive way of obtaining surveillance photographs without alerting the suspect. Surveillance scopes can detect activities at a much lower light level and allow photography of a suspect.

Camera Position

Choice of the camera position can be critical to obtaining useful surveillance photographs. The camera must be positioned to get a useful image size on the film. Where the camera cannot be located near the subject, telephoto lenses must be chosen to provide the needed magnification. Usually, concealment of the camera or cameraman is required so that the subject is not warned that he is being photographed.

When choosing the camera position, try to evaluate the effect that shutter noise, the red glow of infrared flash, the movement of the photographer, or light reflections from the lens will have on the subject. If any of these factors could cause the subject's attention to be attracted, change the camera position, conceal the camera or photographer more effectively, or choose an alternative photographic technique.

Although having the camera at subject's eye level will usually yield the best pictures for identification, this is frequently impractical if the photographer is to avoid having his view obstructed. Try not to place the camera too high, however, since the subject's features may be obscured.

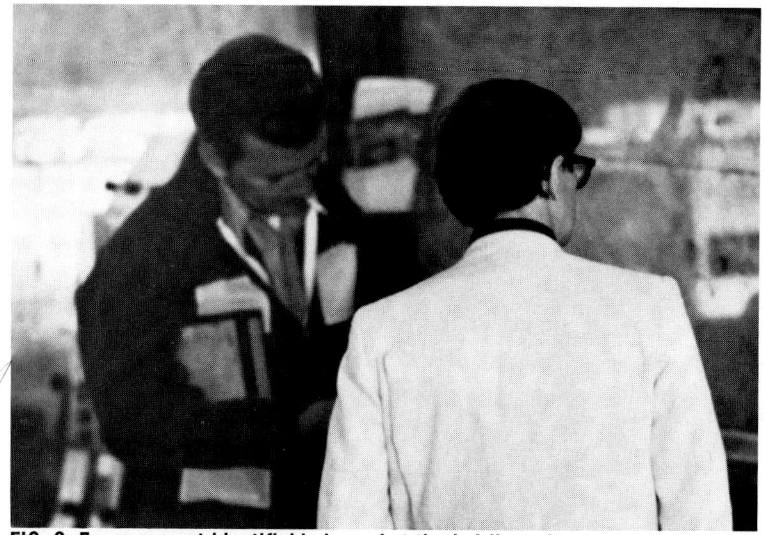

FIG. 2: Faces are not identifiable here, but the hairline, glasses, or clothing may help to establish identity.

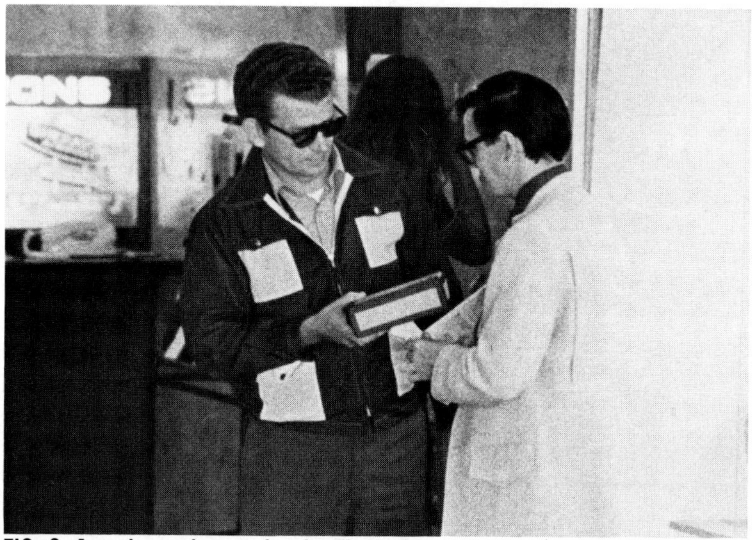

FIG. 3: A package changes hands. The action is recorded and the subjects can be identified.

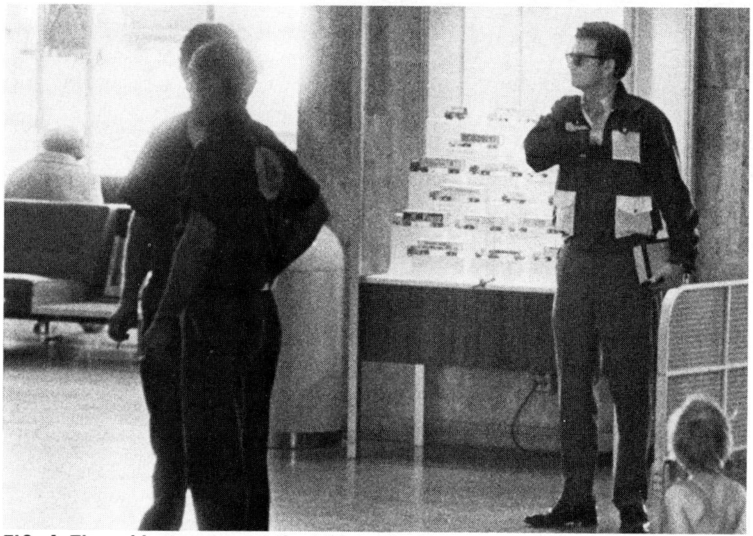

FIG. 4: The subject pauses at the lobby entrance.

FIG. 5: Talking on the phone, the subject is off guard.

FIG. 6: The apparently innocent traveler in the foreground turned out to be the subject's contact.

FIG. 7: The camera position was chosen to cover an airport waiting area. The photographer is concealed by two advertising display cases.

FIG. 8: The suspect appears to see the photographer. The surveillance photographer should freeze. Sudden movement will only alert the suspect. If the photographer has taken care to conceal his position, he is not likely to be detected.

Distance

The distance from the camera position to the subject will dictate the choice of camera lens. The lens must provide sufficient magnification to obtain adequate image size on the negative. Where the normal lens cannot provide the needed image magnification, a telephoto lens must be substituted.

Distance does, of course, lessen the problems associated with concealing the camera position or photographer. In riot situations, it also means that the physical safety of photographer and equipment can be preserved.

Dust, smoke, fog, or heat waves between the subject and the camera can degrade the photograph or make photography impossible. It is also very difficult to track and photograph a subject with a telephoto lens if the subject is moving rapidly or erratically.

Equipment

When the conditions of subject position, camera position, lighting, and distance have been established, you can begin to consider what type of equipment will serve to fulfill the picture requirements. For still photography, the most versatile camera system is a 35mm camera with interchangeable lenses. Fast lenses of all focal lengths allow the camera to be adapted for almost any situation.

Continuous or time-lapse motion pictures can be recorded with 16mm or super 8 cameras. These cameras are capable of operation under a wide range of lighting conditions with interchangeable lenses or zoom lenses. Kodak offers specialized super 8 surveillance cameras which provide a simple and economical technique for monitoring crowds or demonstrations, observing arrests, or detection of crimes.

The widest range of Kodak films, from which a choice can be made to give the required combination of speed and definition, is available in the 16 and 35mm sizes.

The need for photographs taken at very low light levels dictates the choice of equipment and films. Fast lenses and high speed film are necessary to record scenes with low illumination levels.

In addition to the camera system for surveillance, you must consider what accessories will be needed. For telephotography, a firm tripod is a *must.* Proper use and interpretation of an exposure meter can mean the difference between success and failure in many surveillance situations. Supplementary lighting will usually involve infrared lamps or electronic flash equipped with infrared filters.

FIG. 9 & 10: The photographer should steady his camera on any convenient support when he cannot set up a tripod. To avoid attracting attention, the photographer (in the lower photo) preaimed and prefocused his camera on an open door. He then partially concealed the camera with a newspaper and squeezed off the release without bringing the camera to his eye.

Films

Films for surveillance are either moderate- to high-speed panchromatic materials or special-purpose films sensitive to infrared radiation. To achieve the required subject identification, the photographer must choose the film which will yield good definition with speed adequate for the lighting conditions at hand.

Film definition is influenced by the graininess and resolving power of the emulsion. The sharpness of a film image is determined by the quality of the image reaching the film, the degree of exposure, and the method of development.

PHOTOGRAPHIC SURVEILLANCE TECHNIQUES

Existing-Light Photography

The principal challenge for a photographer doing surveillance work is achievement of proper exposure under extreme variations of lighting. Because supplemental light could arouse the suspicion of a subject and because it is frequently just not possible to supplement the scene lighting, much surveillance photography is by existing light. Existing light is, of course, any natural light or artificial light available at the scene.

For low-light-level versatility, you must choose a camera lens with large maximum aperture. Normal lenses for 35mm cameras, with focal lengths from 44mm to 58mm, are usually quite fast—at least *f*/2—and lenses with even larger maximum apertures are common. Because of design limitations, telephoto lenses cannot be made with such wide apertures, but moderately long focal-length lenses (100–300mm) are available with apertures of *f*/2.8 to *f*/4. A wide choice of high-speed films is available in 35mm and 16mm sizes. (Films for 16mm cameras and the use of super 8 cameras and films for low-light photography are described in the section Motion-Picture Photography.)

Light Meters

For measurement of very low light levels, a light meter incorporating a highly sensitive element, such as a cadmium sulfide (CdS) cell, is indispensable. Where the brightness of only a small area of the scene is to be measured, as in surveillance photography, a *spot meter* can be used from the camera position. This type of meter is especially useful for telephoto pictures that include only a small portion of the scene and for photographing scenes with very wide brightness differences.

Some cameras have through-the-lens exposure meters that work on the same principle as spot meters—they read only a small segment of the scene. Other through-the-lens meters read the entire picture area or make a "weighted" reading, with the center spot having greatest influence on the exposure value. Total-area and weighted-area metering systems are less useful for surveillance photography under low-light conditions.

Exposure Indexes

The exposure index assigned to each film is for use with meters and other exposure computing devices which are marked for ASA speeds. The exposure indexes are arithmetic; they indicate directly the relative sensitivity of the various films to light. For example, if all other factors in a system are equal, a film with an exposure index of 200 requires twice as much exposure as one with a speed of 400.

Use of assigned exposure indexes generally leads to the minimum exposure needed to yield good quality negatives. However, if highlight areas of a scene are of prime interest, the effective speed of a film may be increased significantly (at the expense of increased grain and less shadow detail) by developing the film to a higher contrast and printing onto a paper of reduced contrast. See *Push-Processing KODAK Black-and-White Films,* page 18, for more information on achieving higher effective film speeds. Wherever conditions permit, use the lower index since this will permit complete exposure of the film and allow maximum recording of information. Because conditions of exposure in surveillance work vary greatly, use the exposure indexes as a guide for making trial exposures.

FIG. 11: Distant street lamps, low-wattage tungsten lamps, and late evening twilight combine to give this scene adequate illumination. The photographer was careful to avoid underexposure of the scene indicated by his light meter when it was pointed toward the many light sources.

FIG. 12&13: Both scenes were taken under street lighting. The photographer was concealed across the street at a distance of about 100 feet. The 35mm camera was tripod mounted and equipped with a 135mm telephoto lens.

FIG. 12: Taken with KODAK TRI-X Pan Film.

FIG. 13: Taken with KODAK 2475 Recording Film (ESTAR-AH Base).

FIG. 14: Low light level of the bar did not prevent the photographer from recording this scene with KODAK 2475 Recording Film rated at EI 3200. Compare this photograph with the one recording the same scene on infrared film (see page 15, Fig. 27).

Suggested Trial Exposures for Existing-Light Photography

Lighting Conditions	Light Level (footcandles)	Suggested Trial Exposure		
		Kodak 2475 Recording Film rated at EI 3200 and developed* 8 min. in fresh Kodak Developer DK-50	**Kodak 2475 Recording Film rated at EI 1000 and developed* 5 min. in fresh Kodak Developer DK-50** **Kodak Tri-X Pan Film rated at EI 1250 and developed* 11 min. in fresh Kodak Developer D-76**	**Kodak Tri-X Pan Film rated at EI 400 and developed* 7 min. in fresh Kodak Developer D-76**
Light of match, cigarette lighter.	1-2	1/30 sec *f*/2.8	1/30 sec *f*/1.4 ▼ *f*/2	
Nightclub, dim street lighting.	2-4	1/30 sec *f*/4	1/30 sec *f*/2 ▼ *f*/2.8	
Building interiors, corridors, average street lighting.	4-8	1/60 sec. *f*/4	1/30 sec *f*/2.8 ▼ *f*/4	1/30 sec *f*/2
Bus station, sports arenas, stage lighting.	16-64	1/125 sec *f*/5.6	1/60 sec *f*/4 ▼ *f*/5.6	1/30 sec *f*/4
Work areas, store interiors.	32-125	1/125 sec *f*/8	1/60 sec *f*/5.6 ▼ *f*/8	1/60 sec *f*/4
Daylight (heavy overcast).	400-1600		1/250 sec *f*/11	1/250 sec *f*/8 ▼ *f*/11
Bright or hazy sunlight.	5000-10,000		1/500 sec *f*/32	1/250 sec *f*/22 ▼ *f*/32

*At 68 F (20 C) with continuous agitation.

Note: The symbol ▼ indicates a lens opening between the two settings.

KODAK Still Camera Films for Surveillance Photography

KODAK Film	Normal Exposure Index	Trial Exposure Index for Surveillance	Identification or Specification Code	Sizes
TRI-X Pan	400	to 1600	TX	135-20 or 36 exp. 120 rolls 126-12 exp.
TRI-X Pan Professional	320	to 1250	TXP	120 and 220 rolls
TRI-X Pan Professional, 4164 (ESTAR Thick Base)	320	to 1250	—	sheet sizes
2475 Recording (ESTAR-AH Base)	1600[1]	to 4000	RE Sp417	135-36 exp. 35mm x 125 feet[2]
ROYAL-X Pan	1250	to 4000	RX	120 roll
ROYAL-X Pan, 4166 (ESTAR Thick Base)	1250	to 4000	—	4 x 5-inch sheets
High Speed Infrared	200[3]	—	HIE	135-20 exp.
High Speed Infrared, 2481 (ESTAR Base)	200[3]	—	HIE417 or HIE421	35mm x 150 feet[2]
High Speed Infrared, 4143 (ESTAR Thick Base)	200[3]	—	—	4 x 5-inch sheets

[1]Practical index for tungsten light.

[2]If the film is used to reload a 35mm magazine, the magazine should be loaded, stored, placed in the camera, and removed from the camera in total darkness. No safelight should be used. These precautions are necessary because customer-loaded magazines are not sufficiently lighttight to protect high-speed films against accidental exposure.

[3]Tungsten index with no camera filter. See section on **Infrared Photography.**

Telephotography

In telephotography many factors make the task difficult. Adding the requirement that photographs be taken at very low light levels greatly increases the difficulties. As we will explain, telephotography at some distance requires that the photographer use high shutter speeds to avoid image blur and small apertures to provide adequate depth of field. To cope with these requirements and to get pictures at low light levels, the surveillance photographer must use high-speed films.

FIG. 15: 85mm

FIG. 16: 180mm

FIG. 17: 300mm

FIG. 15, 16, & 17: Field of view for lenses mounted on a 35mm camera. The girl in the car is 300 feet from the camera.

Telephoto Lenses

A telephoto lens has a focal length that is longer than the focal length of the camera's normal lens. Technically, the term "telephoto" refers to a particular kind of optical arrangement that has a positive front element and a negative rear element. This allows the physical length of some telephoto lenses to be shorter than the focal length. However, it has become common practice to call any lens with a focal length that is longer than normal a "telephoto" lens.

Telephoto lenses include a narrower angle of view than the normal lens, so they take in a smaller area of the scene. Consequently, distant subjects photographed through a telephoto lens appear closer than they do when they are photographed through a normal lens.

To get adequate image size on the negative, do not exceed the distance shown for each lens focal length:

Lens Focal Length (mm)	Surveillance Distance (feet)
50	35
85-100	60
135	90
180-200	125
300	200
500	330

Depth of Field

Telephoto lenses have shallow depth of field. The longer the focal length of the lens, the more shallow the depth of field. This means that accurate focusing is much more important with telephoto lenses than it is with normal and wide-angle lenses.

From a practical point of view, depth of field varies with the size of the lens opening, the distance of the subject focused upon, and the focal length of the lens. Depth of field becomes smaller as—

The size of the lens opening increases.
The subject distance decreases.
The focal length of the lens increases (and subject distance remains unchanged).

Shutter Speed

The telephoto lens is demanding. In addition to focusing very carefully, you must hold your camera extremely steady to get pictures that are free of blur. This is because a telephoto lens magnifies camera movement as well as image size. Consequently, it's a good idea to use a tripod and a cable release with a telephoto lens. As a general rule, don't try to hand-hold the camera when you're taking pictures with a lens that has a focal length longer than 135mm. If you must hand-hold the camera, use as a minimum speed the shutter speed you get by making the focal length of the lens, in millimeters, a fraction, i.e., for a 50mm lens—1/50 second, 125mm lens—1/125 second, 250mm lens—1/250 second, or 500mm lens—1/500 second.

A telephoto lens increases the effect of subject movement, too. When the subject distance remains the same, the effect of subject movement increases in direct proportion to the focal length of the lens. For example, if you need a shutter speed of 1/250 second to stop the action with a 50mm lens, you'll have to use a shutter speed that's twice as fast, or 1/500 second, with a 100mm lens. The longer the focal length, the faster the shutter speed that will be necessary.

FIG. 18: The center of the negative shot with a 300mm lens was enlarged to give a 1-inch head size on the print. The subject is identifiable.

Zoom Lenses

Zoom lenses, available for some cameras, have a variable focal length. A zoom lens may have a typical focal length variation of 45mm to 135mm, or 85mm to 210mm. The zoom lens has the advantage of offering an infinite selection of focal lengths within its zoom range. Most zoom lenses are in the long-focal-length range, and the information given for telephoto lenses also applies to long-focal-length zoom lenses.

FIG. 19: Even slight camera movement was enough to cause blur with a 300mm lens.

Accessories

With telephoto lenses, use a lens hood to reduce flare and to prevent extraneous light from entering the lens. This is especially important where there are strong sources of light near the subject. Many telephoto lenses have built-in lens hoods; however, if you are choosing a supplementary lens hood, make certain that it does not vignette the image (block light to the corners of the film frame). Vignetting is obvious when you view through the eyepiece of a single-lens reflex camera.

It is often suggested that the simplest method for increasing the image size on the film is interposing a teleconverter, or focal length doubler, between an interchangeable lens and the camera body. Such converters add an additional lens group to the existing lens and they do, indeed, increase the image size. This increase is not without penalties however. Such supplementary lenses usually result in some loss of image quality. In addition, the maximum aperture of the lens combination is severely reduced. Where, for instance, a 2X converter is added to an *f*/4, 200mm lens, the lens plus converter, now 400mm, has a maximum aperture of *f*/8.

FIG. 20: Panning, or moving the camera in the same direction as the subject movement, provided a sharp image of the car and passengers while blurring the background.

Motion-Picture Photography

Much of the information given for surveillance with still cameras is equally applicable to continuous recording of activities with motion picture cameras. Professional 16mm motion picture cameras can be equipped with any of several high-speed, long-focal-length lenses. High speed pan and infrared films are available for these 16mm cameras.

Surveillance with Super 8 Movie Cameras

Many specialized techniques of surveillance are too expensive, complicated, or time-consuming for the average police department. Low light level versatility and relatively modest expenses are involved in using one of the Kodak super 8 cameras designed with active or passive surveillance capabilities.

In active surveillance under extremely low light levels, the KODAK XL55 Movie Camera makes a continuous record of suspicious activities possible. The camera has a fast, *f*/1.2 zoom lens; a 230-degree shutter to deliver maximum available light to each frame; framing rates of 9 and 18 frames per second to give shutter speed control; and the capability of using high-speed super 8 films.

FIG. 21, 22, 23, & 24: A sequence showing subject, locale, and actions may be recorded with a motorized 35mm camera or a motion picture camera. These pictures were taken under extreme conditions: low light, heavy rain, and long distance. Although the conditions limited the quality of the image recorded with KODAK 2475 Recording Film rated at EI 3200, the automobile number plate is readily identifiable in the original.

FIG. 25: This police officer takes advantage of the low light capabilities of a KODAK XL55 Movie Camera to record activities in an underground parking garage.

With the $f/1.2$ lens, actual shutter speeds of 1/28 or 1/14 second, and films rated at an exposure index of 160, this camera is capable of making exposures with 2 to 5 footcandles of illumination on the subject. This capability essentially duplicates the results that can be obtained with most 16mm cameras and KODAK 2475 Recording Film.

Passive Surveillance with Super 8 Movie Cameras

The development of super 8 photographic equipment has made possible a simple, effective, and economical technique for maintaining a passive photographic surveillance system for detection of crime, for observing crowds or demonstrations, and for recording arrests. In some applications, super 8 equipment will complement existing 16mm or 35mm equipment. In other applications, the simplicity and low cost will enable photographic surveillance to be used where it was not feasible before.

Passive surveillance cameras fall into two categories, demand cameras and continuous-surveillance cameras. Demand cameras remain inactive until they are triggered by a switch or other actuating device. The camera then runs at either a continuous frame rate (18 frames per second) or at an intermittent rate, such as one frame every two seconds. Continuous-surveillance cameras are used in situations where a photographic record of movement or events is required over a long period of time (usually hours). This type of camera photographs the scene at an intermittent rate (such as one frame every minute) that compresses time and provides a rapid review of the activity when the processed film is viewed at a normal rate. This type of photography is known as *time-lapse* photography.

The KODAK MONITOR Super 8 Camera is a *demand* camera, designed for shooting 50-foot cartridges of KODAK MFA Film at 18 frames per second. The MONITOR Super 8 Camera is equipped with an $f/1.9$ KODAK Zoom Lens that has a focal length variable from 13mm to 28mm. It covers the full range from wide angle through telephoto. The zoom lens makes it easy to obtain the desired coverage of an area with little difficulty.

The camera can be started and stopped by means of the switch on the KODAK Remote Control Cord or by other external triggering devices. This cord allows the camera to be operated from a remote position.

The KODAK ANALYST Super 8 Camera is designed for *time-lapse* photography, using KODAK MFX Film. This camera exposes film at continuously variable rates from one frame every 1¼ seconds to one frame every 90 seconds. The ANALYST Super 8 Camera is equipped with the same $f/1.9$, 13mm to 28mm KODAK Zoom Lens that is supplied on the MONITOR Super 8 Camera.

Because the KODAK ANALYST Super 8 Camera is a *time-lapse* camera, it is necessary to determine what frame rate is sufficient to do a particular job. For example, there are 72 frames per foot of super 8 film. Therefore, a 100-foot cartridge of KODAK MFX Film contains 7200 frames. To find the running time in seconds for a 100-foot cartridge, multiply 7200 by the time in seconds between frames.

The shutter speed of the KODAK MONITOR Super 8 Camera is 1/40 second, and the shutter speed of the KODAK ANALYST Super 8 Camera is 1/100 second. The exposure index of KODAK MFA Film and KODAK MFX Film in tungsten light is 160.

Viewing Super 8 Surveillance Films

The KODAK EKTAGRAPHIC MFS-8 Projector is an extremely versatile projector for super 8 film. It will show films in either motion or still modes and can be operated at 6, 18, or 54 frames per second. A remote-control feature allows full control of forward and reverse projection, and a simple flip of the switch stops the film where you want. A continuous series of frames can be viewed by repeated actuations of the start button.

The EKTAGRAPHIC MFS-8 Projector provides an excellent way to project surveillance films. A rapid frame rate can be used to run through footage that is of no interest. Footage to be studied closely can then be projected at 6 frames per second; or if close examination is required, the projector can be stopped or advanced frame by frame.

FIG. 26. A KODAK ANALYST Super 8 Camera, mounted in the open rear window of a station wagon, attracts little attention. If it is necessary, the camera can be camouflaged by surrounding it with sports equipment or packing boxes. It may even be inserted into a box with an opening for the lens.

KODAK 16mm Motion Picture Films for Surveillance Photography

Film	Exposure Index	
	Daylight	Tungsten
KODAK TRI-X Reversal Film 7278	200*	160*
KODAK 4-X Reversal Film 7277	400*	320*
EASTMAN DOUBLE-X Negative Film 7222	250	200
EASTMAN 4-X Negative Film 7224	500	400
KODAK 2475 Recording Film (ESTAR-AH Base)	1000	1250
KODAK High Speed Infrared Film 2481 (ESTAR Base)	—	200†

*When reversal processed as recommended.
†No camera filter. See section **Infrared Photography.**

KODAK Super 8 Films for Surveillance Photography

KODAK Film	Exposure Index*	
	Daylight	Tungsten
TRI-X Reversal Film 7278	200	160
4-X Reversal Film 7277	400	320
MFA Film†	200	160
MFX Film‡	200	160

*When reversal processed as recommended.

†KODAK MFA Film is a high-speed panchromatic film in a super 8 cartridge that contains 50 feet of film. It is designed for use in the KODAK MONITOR Super 8 Camera. It can be reversal processed to give a positive, or it can be processed to a negative.

‡KODAK MFX Film is a high-speed panchromatic film designed for use in the KODAK ANALYST Super 8 Camera. Each super 8 cartridge of KODAK MFX Film contains 100 feet of film on ESTAR Thin Base.

Infrared Photography

Photographic materials can be sensitized to record radiation which is not visible to the human eye. This fact is of great importance for surveillance photography. If an area is illuminated with infrared radiation, a suspect can be observed and photographed with little likelihood of alerting him.

In relatively confined areas, suitably filtered*, continuously burning tungsten lamps will usually provide all of the infrared radiation needed for either still- or motion-picture photography with KODAK High Speed Infrared Film. In places such as theaters, nightclubs, or bars, where there is already some visible light, supplemental infrared lamps will allow still- or motion-picture records to be made. An observer may see a dim red glow from an infrared lamp if he looks directly at it. He will not notice it if the lamp appears to be part of an exit light, a neon sign, or other display lighting.

KODAK High Speed Infrared Film is often used in law-enforcement surveillance photography to obtain a photographic record under unfavorable lighting conditions, without introducing supplementary lighting. The exposures given below are very much lower than those required for results of optimum quality and are just adequate to render details barely recognizable.

*KODAK WRATTEN Filter No. 87 or KODAK Safelight Filter No. 11. See Kodak Publication No. M-28, *Applied Infrared Photography,* for details.

1. Basic exposure with visible tungsten radiation measuring 0.5 to 1.0 footcandle incident at the subject: *f*/1.9 at 1/30 second or 16 frames per second. With no filter, but with recommended development.
2. Low visible radiation, high infrared radiation: Replace the tungsten lamps in No. 1 above with an equal wattage of clear ruby lamps. Exposure and development conditions are the same as above.
3. At a lamp-to-subject distance of about 40 feet and using one 250-watt GE Reflector Infrared Heat Lamp (250R40/10) with with ruby glass: *f*/1.9 at 1/25 second with recommended development.

Infrared Flash

Where an instantaneous infrared source is needed, special flash bulbs or an electronic flash covered with an infrared-transmitting filter will provide the required radiation. Cover electronic flash units with a KODAK WRATTEN Filter No. 87 or No. 87C. These filters will absorb practically all of the visible radiation so that, even in a completely darkened room, the intruder will not be conscious of any visible light, unless he is looking directly at the flash. Then he will see only a very faint red glow. To eliminate this possibility, the flash reflector can be pointed towards the ceiling and the picture made by reflected infrared radiation, in the same way that bounce light is used in regular flash photography. This bounce-light technique is most effective if ceiling height is no greater than 10 feet.

Because the lamp does not have to be replaced after each exposure with electronic flash units, the filter can be taped over the reflector or held in place with rubber bands. (Make sure no stray light is visible around the filter.) The variables

FIG. 27: This covert photograph was made with infrared radiation from an electronic flash covered with a KODAK WRATTEN No. 87 Filter. The flash was pointed upward to bounce the infrared radiation off the wall and ceiling. No filter was used on the camera which was loaded with KODAK High Speed Infrared Film. Note the change in the appearance of clothing as compared with Fig. 14, page 7.

in the use of infrared photography are so numerous that specific exposure settings should be determined by tests. If you do not know the beam-candle-power-second (BCPS) output of your electronic flash, try a guide number of about ¼ to ⅓ that which you use for Tri-X Pan Film. With an electronic flash unit of 1000 BCPS or ECPS, a Kodak Wratten Filter No. 87 and Kodak High Speed Infrared Film, start exposure tests with a guide number of about 55. With a 4000 BCPS unit try a guide number of 110 as a starting point.

Focusing for Infrared

Infrared rays, because of their longer wavelength, do not focus in the same plane as visible rays. It is therefore necessary to make an increase in the lens-to-film distance to correct for the focusing difference between infrared and visible rays. For best definition, infrared photographs should be exposed with the smallest lens opening that conditions permit. If large apertures must be used, and the lens has no auxiliary infrared focusing mark (usually a small red dot), a focusing correction can be established by photographic focusing tests. A basis for trial is the *extension* of the lens by ¼ of 1 percent of its focal length or a shift in the focus setting to a *nearer* distance by about the space between the infinity and 50-foot marks.

Infrared Surveillance Scopes

Several manufacturers provide infrared surveillance scopes that allow continuous observation of a suspect without detection. These devices use an image converter to translate the infrared into a visible image which is displayed on a screen. The screen is observed through an eyepiece and can also be photographed through this eyepiece. The infrared scope can be used to direct a camera which can photograph the suspect with the ambient infrared radiation or with infrared flash.

Using a Light Meter as a Guide for Exposures on Infrared Film

Regular exposure meters, which are essentially sensitive to visible radiation only, cannot be used with infrared film in daylight because of the varying ratio of infrared to visible radiation. With tungsten light sources, however, meters can be used as indicators of exposure. For infrared film, the exposure is best determined by actual photographic tests under prevailing conditions.

Light-Intensifier Scopes

Even in relatively dim light, high-speed films can usually capture the subject if the photographer can see clearly enough to focus his camera. For still lower light levels, developments in electronics provide a means for seeing and photographing subjects where even the eye cannot discriminate objects. Unlike the devices that convert invisible infrared radiation to visible light, these image intensifier systems require no external radiation source. Instead, the light provided by a distant street light, moonlight, or even starlight and reflected by the subject is captured and amplified electronically. The output of the electronic amplifier is converted to a bright image which can be observed with the eye or which can be photographed.

Most image intensifier systems have a high-speed, long-focal-length lens which forms the low-light-level image on an image converter tube. This tube converts the light image to electronic signals which are amplified several thousand times. The signal is transmitted to a screen that produces a visible display of the amplified image.

Observation and photography of the display can be done with a 35mm single lens reflex camera or a reflex-viewing movie camera placed at the eyepiece. (The manufacturer of the image intensifier unit recommends appropriate extension tubes or other devices required to focus the camera properly.) Adjusting the intensity of the displayed image may be done with the aperture of the objective lens on the scope. Use Kodak Tri-X Film or Kodak 2475 Recording Film (with normal processing) to photograph the display. The scope manufacturer provides a guide for trial exposures. Do not rely on through-the-lens metering systems to establish exposure requirements. The meter may serve as a guide to exposure after tests have been made to relate its response to actual film exposure.

Low-light image intensifier mechanisms can also be coupled to closed-circuit-television monitoring systems. Such a display can be monitored by several persons and photographed simultaneously. To photograph a closed-circuit-television monitor, load the camera with Kodak Tri-X Pan Film, focus the camera on the screen, and shoot at 1/30 second and *f*/5.6 with a leaf-type shutter or at ⅛ second and *f*/11 with a focal-plane shutter.

FIG. 28: Over the face of the intensifier tube, shown here, a focusing lens and camera back are fitted. Through a reflex prism, the photographer can observe actions which can be recorded on film by actuating the camera's focal-plane shutter.

Passive Surveillance

Normally in police surveillance, the camera is actively manned by an officer or controlled by him with a remote switching device. There are times, however, when a passive surveillance system, triggered by the suspect himself, can serve as a photographic stakeout. Many devices can sense the presence of an intruder and trigger the operation of a camera.

Photoelectric Detectors

Photoelectric detectors are among the oldest and most widely used types of detectors. One type of photoelectric detector relies on a beam of light falling on a photosensitive detector. When the light beam is broken, by a person entering a room for instance, the interruption causes the camera to operate.

Another type of photoelectric unit relies on the use of a detector that is sensitive to infrared radiation. The detector senses an increase in infrared radiation such as that radiated from the body of an intruder, and through electronic circuitry, causes the camera to operate.

Audio Detectors

This system relies on a microphone detector that detects only unusual or out-of-the-ordinary sounds such as those made by a person moving items in the protected area.

Vibration Detectors

These are very sensitive detectors of any movement in the protected area. They are generally employed in areas in which there should be no personnel during particular hours.

Capacitance Detectors

A capacitance field is established around the area. A person entering the area disturbs the balance of this field; this imbalance is detected by the unit and the camera is actuated.

Ultrasonic Detectors

The area is filled with ultrasonic sound waves having a frequency above 20,000 hertz. A frequency this high cannot be heard by the human ear. However, movement by a person in the area will cause a disturbance of the sound-wave pattern monitored by the detector, and the detector will activate the camera.

HANDLING AND PROCESSING SURVEILLANCE PHOTOGRAPHS

Storage and Handling of Special-Purpose Films

Films designed to have high sensitivity to red and infrared radiation are also sensitive to the effects of heat and humidity. KODAK 2475 Recording Film and KODAK High Speed Infrared Films require special care in handling and storage.

Whenever possible, store unexposed recording films (in sealed containers) in a refrigerator or freezer. Unexposed rolls of KODAK High Speed Infrared Film *must* be stored in a refrigerator or freezer at 55 F (13 C), or lower, in the original container. Where the unexposed films are to be kept for extended periods of time (6 months or more), store film in sealed packages at 0 to −10 F (−18 to −23 C). If the film is stored in a refrigerator, remove it 2 hours before opening the package or can; if stored in a freezer, remove it about 4 hours before opening. This will prevent condensation of atmospheric moisture on the cold film.

Keep exposed film cool and dry. Process the film as soon as possible after exposure to avoid undesirable changes in the latent image. If it is necessary to hold exposed but unprocessed film for several days (such as over a weekend) it should be refrigerated below 40 F (4 C).

Infrared films *must* be handled in *total* darkness. This includes camera loading and unloading as well as processing. When processing infrared films, you should make sure that your darkroom does not admit infrared radiation. Some types of film developing tanks may also leak infrared radiation.

To avoid fogging recording films, load and unload the camera under subdued light or, better still, in total darkness. Make certain that the darkroom is actually lighttight for loading processing reels or loading cassettes from bulk rolls. Keep the film in its sealed container until it is loaded into the camera, and return it to the container after exposure to await processing.

Processing Surveillance Films

Where exposures are made under normal lighting conditions, medium-contrast, fine-grain development will provide the most useful negatives. In many cases, however, the lighting conditions are not ideal, so that films must be exposed and processed for maximum film speed. It is best to choose a low-light-level surveillance film which has recommendations for processing to obtain maximum speed. Where such a film is not available, however, it is possible to increase the speed rating of normal black-and-white films by push-processing.

Push-Processing KODAK Black-and-White Films

For surveillance situations in which you don't have enough film speed and cannot change to a high speed surveillance film, you may have to underexpose and push-process the film. Although some quality is lost when you underexpose and push-process your black-and-white film, you can accept the slight loss in detail and image quality to get a picture that would otherwise be impossible. By increasing development time, you can obtain acceptable quality in *most* black-and-white pictures that have been underexposed by 1½ to 2 stops. The amount by which you can underexpose will depend on the contrast of the scene. For example, you can expose KODAK TRI-X Pan Film at EI 1250-1600 instead of its normal speed of EI 400.

Normal film speeds are based upon adequate exposure in the shadow end of the tone scale. Increasing development does not greatly increase the film speed based upon this shadow rendition. However, increased development does increase the contrast of most black-and-white films, and it significantly increases the densities in the middle-tone and highlight areas. There is, in effect, a speed increase for the middle-tone and highlight areas of the negative. This means that the film can be rated at a much higher than normal speed (with appropriate development) if the photographer is willing to sacrifice shadow detail.

Negatives that are underexposed and overdeveloped can give usable prints at the expense of increased grain and loss of shadow detail. Underexposed negatives will show better quality with push-processing than with normal processing, but they will not produce the same quality prints as normally exposed and processed negatives would. The prints will probably have poor shadow detail, but they should show good detail in the highlights and middle tones which are important in surveillance.

To push-process your black-and-white film when you have underexposed it by 1½ to 2 stops, increase the development time recommended on the film instruction sheet by 50 percent. This is only a general rule, because the limit for push-processing varies according to the film and developer combination you use. For push-processing Kodak films, you can use any of the developers recommended on the film instruction sheets, except KODAK MICRODOL-X Developer.

Processing Films for *Maximum* Speed

To achieve maximum film speed, follow the processing recommendations very carefully. KODAK 2475 Recording Film can be processed for maximum film speed at normal temperatures in KODAK Developer DK-50. KODAK TRI-X Pan Film gives best speed and sharpness characteristics when developed in KODAK Developer D-76.

Handle KODAK 2475 Recording Film in total darkness. Do not attempt to develop by inspection. Turn on room lights or open the developing tank only after the film is fully fixed. KODAK TRI-X Pan Film should also be handled in total darkness. However, when development is half-completed, you can use a safelight at 4 feet for a *few* seconds. Equip the safelight with a KODAK Safelight Filter No. 3 (dark green) and a 15-watt bulb.

To open the 135 magazine, hold the magazine with the long end of the spool down and use a lid lifter or hook-type bottle opener to remove the upper end cap from the magazine. Kodak 135 magazines have hard-staked end caps; they are not intended for reuse. Do *not* attempt to save and reuse magazines that are not especially designed for reloading.

Processing Films for *Maximum* Speed

Develop: With continuous agitation.

KODAK Film	KODAK Developer	Time (in minutes)	Temperature
2475 Recording	DK-50	8 minutes	68 F 20 C
TRI-X Pan	D-76	11 minutes	
ROYAL-X Pan	DK-50	7 minutes	

Rinse: 30 seconds with continuous agitation in KODAK Stop Bath SB-la at a temperature within 5 F of the developer temperature.

Fix: With frequent agitation for 5-10 minutes in KODAK Fixer or KODAK Fixing Bath F-5 at 65-70 F (18-21 C). Or, using KODAK Rapid Fixer, 2-5 minutes at 65-70 F (18-21 C).

Wash: In running water for 20-30 minutes at 65-70 F (18-21 C).

Rapid Wash: To reduce wash time and to conserve water, rinse the film in running water for 20 seconds to remove excess hypo, and then immerse it in KODAK Hypo Clearing Agent solution for 1 to 3 minutes. Follow this with a brief wash allowing at least one complete change of water.

Dry: In a dust-free place.

Processing KODAK High Speed Infrared Films

Process infrared roll films on spiral reels in a closed stainless steel tank. If you wish to process infrared films in closed plastic roll-film or rewind tanks in normal room light, first test the tank to determine if it is opaque to infrared radiation, or carry out the development in total darkness.

Develop: In KODAK Developer D-76 for 12 minutes at 68 F (20 C). Agitate at one-minute intervals during development.

Rinse: In KODAK Indicator Stop Bath or KODAK Stop Bath SB-5 at 65 to 75 F (18 to 24 C) for 30 seconds with agitation.

Fix: In KODAK Fixer of KODAK Fixing Bath F-5 for 5 to 10 minutes at 65 to 75 F (18 to 24 C), or in KODAK Rapid Fixer for 2 to 4 minutes at 65 to 75 F (18 to 24 C). Agitate frequently during fixing.

Wash: For 20 to 30 minutes in running water at 65 to 75 F (18 to 24 C).

Rapid Washing: To conserve water and reduce washing time, rinse the film in clear, running water for 30 seconds to remove excess hypo. Then, immerse the film in KODAK Hypo Clearing Agent solution for 1 to 2 minutes. Follow this with a 5-minute wash, using a water flow sufficient to give at least one complete change of water in 5 minutes.

Dry: In a dust-free area.

Printing Surveillance Photographs

Use the best quality enlarging lens that you can get. The best of negatives will not yield a good, sharp print with a poor quality or dirty enlarging lens. Enlarging lenses differ from picture-taking lenses in that they are designed to work best at short lens-to-subject distances. In enlarging, the negative in the subject and the easel is the focal plane.

To make 8-10X enlargements from a full 35 mm negative frame, a 50 mm enlarging lens will be suitable. If you have to enlarge small portions of the 35 mm frame, then a 25 mm enlarger lens may be needed to get sufficient magnification at reasonable working distances. If you expect to make enlargements from super 8 frames, a 12 mm enlarging lens will be necessary to provide enough magnification.

For good definition use an enlarger with an optical-condenser system. The slightest vibration of your enlarger will also cause a loss of definition. A vertical enlarger is prone to vibrate when the head is at or near the top of the column.

Since definition is of prime importance in prints, use a smooth-surface paper such as KODABROMIDE Paper, F (glossy); KODAK POLYCONTRAST Paper, F (glossy); or KODAK EKTAMATIC SC Paper, F (glossy). Use a contrast grade of paper (or, in the case of POLYCONTRAST and EKTAMATIC SC Papers, a POLYCONTRAST Filter) that suits the negative. Apparent sharpness in a print is impaired by using too-low contrast paper. Too-high contrast paper tends to increase the appearance of graininess.

Because of the widely varying lighting conditions existing when surveillance pictures are made, it is seldom that uniformly exposed negatives will result. Contrast of the negative will vary with scene lighting conditions and the extremes to which the film speed must be pushed.

Contrast control can be obtained easily in the printing operation through the use of KODAK POLYCONTRAST or EKTAMATIC SC Papers. These are variable-contrast papers designed to give a wide range of contrasts. The print contrast is controlled by means of filters used between the light source and the printing paper. The KODAK POLYCONTRAST Filter Kit (Model A) can be used with these papers to provide seven different contrasts.

The KODAK EKTAMATIC Processor, Model 214-K, and KODAK EKTAMATIC SC Paper greatly simplify the printing of negatives from surveillance films in almost any location. The processor is a compact, tabletop machine capable of processing an 8 x 10-inch print in 15 seconds. It is a stabilization processor that utilizes activator and stabilizer solutions in place of the conventional developer, stop bath, fixer, and wash.

Negatives made during surveillance assignments may present a problem in printing when the central subject is a contrasty highlighted area surrounded by thin, underexposed shadow areas. A print on Grade 2 paper from this sort of negative must always be a compromise between the shadows and highlights. In using variable-contrast paper, however, you can print the whole negative through a high-contrast filter to give shadow contrast, and then replace this filter with a low-contrast filter to get highlight detail by burning-in the contrasty areas. This method of printing difficult negatives often yields results that cannot be obtained in any other way.

Attempt to retain only as much shadow detail in the print as is relevant to the clear exposition of a subject's actions. It may be difficult to produce a finished print which shows adequate shadow detail and well-defined highlights (such as the subject's face) without resorting to dodging or burning-in. Use these techniques where necessary to show the required information, but be prepared to explain the technique if the photos are used in evidence. In some jurisidictions such manipulations may not be acceptable for courtroom use; however, the prints may be useful for intradepartment analysis and information.

FIG. 29: In the darkroom, the photographer can make quick prints of surveillance negatives on a KODAK EKTAMATIC Processor, Model 214-K.

GLOSSARY

Active surveillance Continuous watch or observation by an officer.

Beam-Candlepower-Seconds The quantity of light obtained from an electronic flash unit is expressed in BCPS (or occasionally "effective-candle-power-seconds, ECPS, which is approximately equivalent). Watt-second ratings are *not* equivalent.

Contrast The density range of a negative, print, or slide; the brightness range of a subject or the scene lighting.

Covert Hidden; secret.

Definition The impression of clarity of detail perceived by an observer viewing a photograph.

Density Degree of darkening caused by exposure and development of a photosensitive material.

Depth of field The distance range between the nearest and farthest objects that appear in acceptably sharp focus in a photograph. For all practical purposes, it depends on the lens opening, the focal length of the lens, and the distance from the lens to the subject.

Existing light Natural light or artificial light unsupplemented by photoflash, photofloods, etc.

Exposure meter An instrument with a light-sensitive cell that measures the light reflected from or falling on a subject; used as an aid to selecting the exposure setting. The same as a light meter.

Exposure setting The lens opening and shutter speed selected to expose the film.

Film speed The sensitivity of a given film to light, indicated by a number; the higher the number, the more sensitive, or "faster," the film.

Focal length The distance from the lens to a point behind the lens where light rays are focused when the distance scale is set on infinity. Focal length determines image size at a given lens-to-subject distance.

Fogging Darkening or discoloring of a negative or print or lightening or discoloring of a slide caused by (1) exposure to nonimage-forming light to which the photographic material is sensitive, (2) too much handling in air during development, (3) overdevelopment, (4) outdated film or paper, or (5) storage of film or paper in a hot, humid place.

Footcandle Unit of measure for illumination; also, lumen/ft^2 (1 footcandle = 10.8 lux).

Graininess	The sand-like or granular appearance of a negative, print, or slide resulting from the clumping of silver grains during development of the film; graininess becomes more pronounced with faster films, increased density in the negative, degree of enlargement, and, generally, increased development.
Infrared	Invisible radiation with wavelength longer than that of visible.
Light-intensifier	A passive (no supplemental light) instrument capable of detecting and amplifying very low levels of light; e.g., "star light."
Normal lens	A lens that makes the image in a photograph appear in a perspective similar to that of the original scene. A normal lens has a shorter focal length and a wider field of view than a telephoto lens and a longer focal length and narrower field of view than a wide-angle lens.
Overt	Open; without concealment.
Passive surveillance	Area monitored by unmanned cameras or camera traps.
Resolution	Ability of a film to record fine detail.
Spot meter	Exposure meter with narrow angle of view allowing light readings from a small area of the subject.
Surveillance	Close watch or observation.
Telephoto lens	A lens that makes a subject appear larger on film than does a normal lens at the same camera-to-subject distance; a telephoto lens has a longer focal length and narrower field of view than a normal lens.
Time-lapse photography	Photography of a scene at intermittent rate (e.g. 1 frame per second, 1 frame per minute) then review at normal rate to compress time.
Underexposure	A condition in which too little light reaches the film, producing a thin negative, a dark slide, or a muddy-looking print.
Wide-angle lens	A lens that has a shorter focal length and a wider field of view (includes more subject area) than a normal lens.
Zone focus	Zone defined by depth of field at a given aperture setting; used where refocusing is impractical or impossible.
Zoom lens	A lens in which the focal length can be adjusted over a wide range, giving the photographer, in effect, lenses of many focal lengths.

FOR MORE INFORMATION

In the Kodak publications listed below, you can find more information on the products and techniques discussed in this book. These publications can be obtained from your photographic products supplier, or from Eastman Kodak Company, Department 454, Rochester, New York 14650. Please list the publication title and code number and include any applicable state and local taxes with your order (minimum order $1). Single copies of those technical pamphlets marked with an asterisk can be obtained, at no cost, from Department 412-L. Please indicate that your request is *from* Kodak Publication No. M-8.

Code Number	Publication Title	List Price
AC-44	Adventures in Existing-Light Photography	$ .95
AC-56	Advanced Camera Techniques for 126 and 35mm Cameras	.95
D-9	Small-Batch Processing of KODAK PLUS-X, TRI-X, and 4-X Reversal Films	.25*
F-4	KODAK Color and B/W Films for Still and Movie Cameras	.10*
F-5	KODAK Professional Black-and-White Films	1.50
G-5	Professional Printing in Black-and-White	1.00
L-5	Index to Kodak Information	.05*
M-28	Applied Infrared Photography	2.00
N-17	KODAK Infrared Films	.75
P-95	KODAK 2475 Recording Film (ESTAR-AH Base)	.10*
P-233	Super 8 Photographic Surveillance	1.00
R-20	KODAK Master Darkroom DATAGUIDE	5.50

*See ordering information above.

List prices shown are suggested prices only and all prices are subject to change without notice.

Suggested Trial Exposures for Existing-Light Photography

Lighting Conditions	Light Level (footcandles)	Suggested Trial Exposure		
		KODAK 2475 Recording Film rated at EI 3200 and developed* 8 min. in fresh KODAK Developer DK-50	KODAK 2475 Recording Film rated at EI 1000 and developed* 5 min. in fresh KODAK Developer DK-50 KODAK TRI-X Pan Film rated at EI 1250 and developed* 11 min. in fresh KODAK Developer D-76	KODAK TRI-X Pan Film rated at EI 400 and developed* 7 min. in fresh KODAK Developer D-76
Light of match, cigarette lighter.	1-2	1/30 sec f/2.8	1/30 sec f/1.4 ▼ f/2	
Nightclub, dim street lighting.	2-4	1/30 sec f/4	1/30 sec f/2 ▼ f/2.8	
Building interiors, corridors, average street lighting.	4-8	1/60 sec. f/4	1/30 sec f/2.8 ▼ f/4	1/30 sec f/2
Bus station, sports arenas, stage lighting.	16-64	1/125 sec f/5.6	1/60 sec f/4 ▼ f/5.6	1/30 sec f/4
Work areas, store interiors.	32-125	1/125 sec f/8	1/60 sec f/5.6 ▼ f/8	1/60 sec f/4
Daylight (heavy overcast).	400-1600		1/250 sec f/11	1/250 sec f/8 ▼ f/11
Bright or hazy sunlight.	5000-10,000		1/500 sec f/32	1/250 sec f/22 ▼ f/32

*At 68 F (20 C) with continuous agitation.

Note: The symbol ▼ indicates a lens opening between the two settings.

Professional and Finishing Markets Division

EASTMAN KODAK COMPANY • ROCHESTER, NEW YORK 14650

Kodak

PHOTOGRAPHIC SURVEILLANCE TECHNIQUES
for Law Enforcement Agencies
Kodak Publication No. **M-8**
CAT 150 6070

10-73-BE Minor Revision
Printed in the United States of America